Housing
230

滑溜溜鳗鱼

Slippery Eels

Gunter Pauli

[比] 冈特·鲍利 著

[哥伦] 凯瑟琳娜·巴赫 绘

靳维筠 译

上海遠東出版社

特别感谢以下热心人士对童书工作的支持：

匡志强　方　芳　宋小华　解　东　厉　云　李　婧
刘　丹　熊彩虹　罗淑怡　旷　婉　杨　荣　刘学振
何圣霖　王必斗　潘林平　熊志强　廖清州　谭燕宁
王　征　白　纯　张林霞　寿颖慧　罗　佳　傅　俊
胡海朋　白永喆　韦小宏　李　杰　欧　亮

目录

Contents

一头格陵兰睡鲨注视着正穿梭于大西洋的成千上万条鳗鱼。他很好奇这些大眼鳗鱼将游去何方。

“你们看起来都很着急呀。”他说。“为什么不慢下来呢？看看周围可以发现新奇的事物，甚至还能遇见一些意外的朋友。”

A Greenland shark is observing thousands of eels travelling through the Atlantic Ocean. He wonders where these big-eyed eels are heading.

“You all seem to be in such a rush,” he comments. “Why not take your time? Look around, discover something new, even meet some unexpected friends.”

一头格陵兰睡鲨注视着成千上万条鳗鱼

A Greenland shark observes thousands of eels

……我的童年将近百年……

... my childhood lasted about a hundred years ...

“我们正赶去公共繁殖区。”一条鳗鱼回应道。“没有多余时间可浪费，因为我们不能让身体里储存的能量耗尽。”

“知道么，我的童年将近百年，我可以轻易地活到四百多岁。所以我给你们的建议是：坚持做一个孩子，能多久就多久。”

“天哪！无论和谁比，你的一生真的好长啊！我们可以活的时间不过你的五分之一，也就大概八十年左右——前提是我们不被捕杀，不用去满足人类的味蕾。”

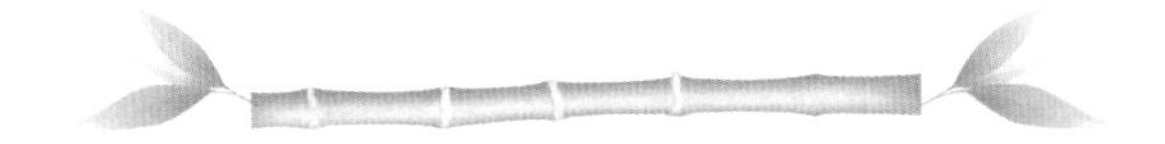

“We are travelling to our common breeding grounds,” an eel replies. “With no time to waste, as we cannot let the energy stored in our bodies run out.”

“Look, my childhood lasted about a hundred years, and I can easily live for up to four hundred years. So my advice is: stay a kid as long as you can.”

“Wow! That is a long life, by any standard. We live for about a fifth of that, about eighty years or so – if we do not get killed to satisfy people’s appetites.”

“我们有一些共同点：像人类一样，我们都是捕食者，我们都吃自己身边的动物，而且我们都可以发电。”

“可是，人类把电鱼也叫作鳗鱼，虽然他们更接近于鲇鱼。我们与他们的共同之处不过就是长得很像而已，以及都被叫作‘鳗鱼’。”

“真的么？我还纳闷为什么人类会把你们的名字弄混呢。”

“We have some things in common: we are both predators, like people, and eat animals around us. And we make electricity.”

“Well, people call electric fish eels too, even though they are closer to catfish. All we have in common with them is the look, and the name ‘eel’.”

“Is that so? I wonder why people get confused with names.”

……人类把电鱼也叫作鳗鱼……

… people call electric fish eels too …

你是准备攻击了么？

Are you getting ready to attack?

“实际上，我们看起来更像是蛇，而不是鱼。当我们在大海中出生时，看起来像是一片扁平的树叶。等我们回到河里，会变成圆柱形，有着大大的嘴巴和小小的牙齿。”

“我看你的嘴巴总是一张一合，那看起来很可怕呢。你是准备攻击了么？”

“才不是呢！大家应该明白，当我开合我的嘴时，我只是在正常地呼吸，没有一丁点威胁他人的意思。”

“In reality, we look more like a snake than a fish. When we are born in the sea, we look like a flat leaf. By the time we return to the rivers, we are cylinder-shaped, with a big mouth and tiny teeth.”

“I see you are always opening and closing your mouth. To many that looks scary. Are you getting ready to attack?”

“Not at all! People should understand that when I open and close my mouth, I am only breathing, and not threatening anyone.”

“但是你滑溜溜的啊！”

“那是。这让任何一个试图抓我的人都难以牢牢地抓住我。我可以穿梭于珊瑚中间，也可以溜进岩洞中而不受一点划伤，而且我皮肤上那厚厚的一层黏液使寄生虫很难侵入我的皮肤。”

“你太聪明了！但我还是疑惑，没有鳍你怎么能游泳？”

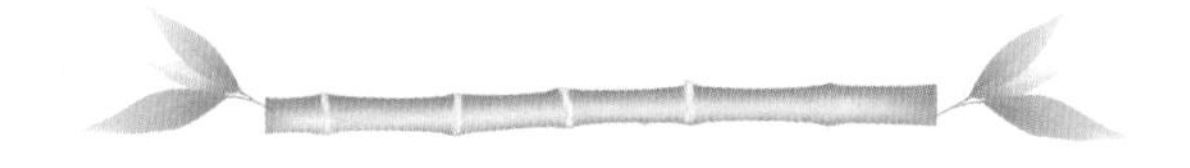

“But you are very slippery!”

“True. Anyone who tries to catch me will struggle to get a grip and hold on. I slide through corals and caves without even suffering a scratch. And this thick slimy layer on my skin makes it hard for parasites to get under my skin.”

“Smart you are! But I do wonder: how can you swim without fins?”

我可以穿梭于珊瑚中间而不受一点划伤

I slide through corals without a scratch

……我们是效率最高的游泳健将。

... we are most efficient swimmers.

“哦，我通过制造水波来移动。”鳗鱼说道。

“但是单靠制造水波无法让你前行吧……”

“当然可以，以我制造水波的方式就可以。我甚至可以通过制造水波来向后移动。”

“这怎么可能？”

“看啊，我们是效率最高的游泳健将。人类根本无法和我们媲美。”

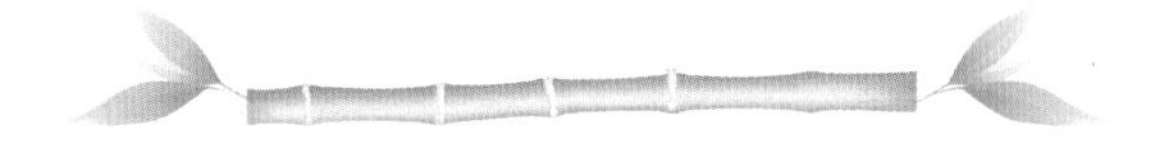

“Oh, I move by making waves,” Eel replies.

“But making waves won’t make you go forward…”

“It does, the way I make waves. I can even choose a wave action that makes me move backward.”

“Now how is that possible?”

“Look, we are most efficient swimmers. People are no match for us.”

“人类创造了很多泳姿，从狗刨式到自由泳。他们模仿青蛙蹬腿的姿势，甚至还学海豚游泳时尾巴摆动的样子。”

“人类坚信想要在水中前行，必须让水流向后运动。”

“有道理。”

“我也觉得这很有道理，但是制造出一种像数字8一样的水波，也许更有道理。”

“People have many swimming styles, from doggy paddle to crawl. They imitate a frog’s kicking, and even the movement of a dolphin’s tail.”

“People believe that in order to move forward, you have to move water backward.”

“Makes sense.”

“I agree that it makes sense, but making waves that create the shape of an eight makes even more sense.”

像数字8一样的水波……

Waves that create the shape of an eight ...

行走？我不行。但是滑来滑去是可以的。

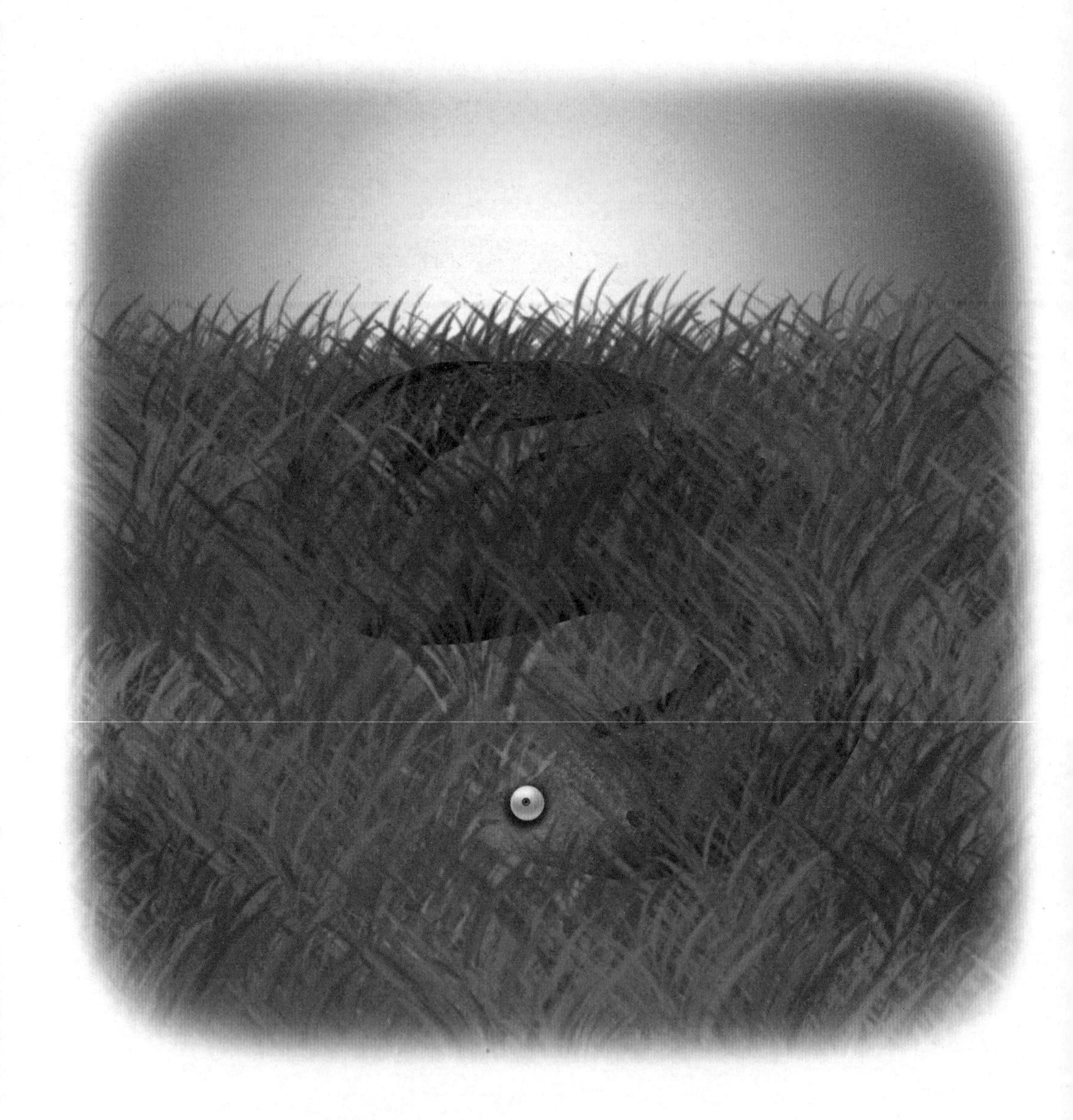

Walk? No. But slide, yes.

“数字8么？我不是很明白。我只是摆动我的尾巴，然后就前行了。”

“但我会摆动全身，将身体一边的水推走，这样水流就将我推向另一边。”

“这真是太有意思了。你真的可以在陆地上行走吗？”

“行走？我不行。但是滑来滑去是可以的。我可以向你保证：当我们鳗鱼下定决心要去某处，没有任何事物可以阻挡我们，即使那意味着要跨越千难万险！”

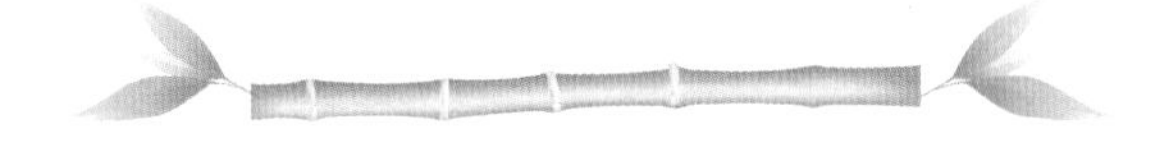

“The number eight? I don’t understand. I just move my tail and I go forward.”

“And I move my whole body, pushing the water away on one side, which pushes me forward on the other side.”

“Interesting … I must say. Is it true that you can walk on land?”

“Walk? No. But slide, yes. I can assure you: once we eels make up our mind to go somewhere, nothing can stop us – even if it means we have to cross obstacles in our thousands!”

“我明白！但是你是如何呼吸的？”

“我们通过皮肤汲取所需的氧气，可以从草地的空气中，也可以从泥巴地、河流或者海洋的水中汲取氧气。”

“你真是一种令人惊叹的物种。我很敬仰你的决心。你很确定自己要什么。”

“比起知道自己要什么，我更清楚自己是谁：一条滑溜溜的鱼，而且以此为傲！”

……这仅仅是开始！……

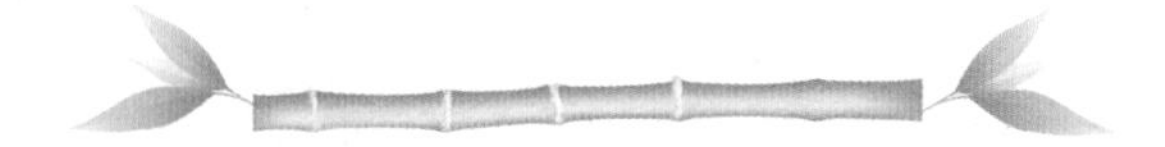

“I can see that! But how do you breathe?”

“We get all the oxygen we need through our skin, from the air when sliding over a meadow, or from the water when in muddy sand, or a river or the sea.”

“You are such a surprising creature. I do admire your determination. You know what you want.”

“And on top of knowing what I want, I know who I am: a slippery fish – and a proud one at that!”

... AND IT HAS ONLY JUST BEGUN!...

……这仅仅是开始！……

... AND IT HAS ONLY JUST BEGUN! ...

Did You Know?

你知道吗?

As eels undulate, pockets of low pressure are created inside each bend of their bodies, into which water is sucked back, allowing the eel to ripple ahead. The force is suction instead of pressure for thrust.

当鳗鱼摆动时，在其身体的每个拐弯处都会形成低压区，水会被吸回去，这能使鳗鱼向前游动。这种动力的来源是吸力，而不是推力。

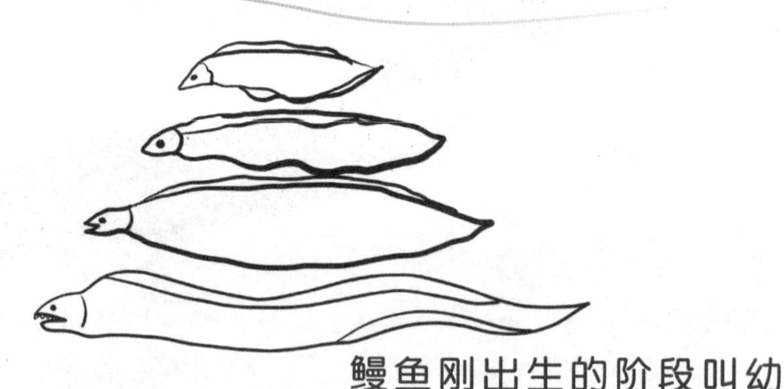

Eels begin life as larvae. Larvae change into glass eels – transparent juvenile eels, and these become elvers. Many eels remain in the sea throughout their lives, but some freshwater elvers travel upstream.

鳗鱼刚出生的阶段叫幼体，幼体成长为玻璃鳗，也就是透明的鳗苗，这些玻璃鳗再长为鳗线。很多鳗鱼的一生都生活在大海里，但是有些淡水鳗线会游行到水的上游。

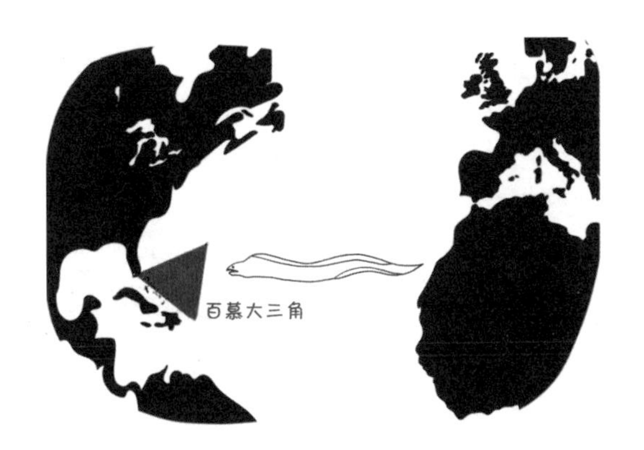

All European eels, from Northern Africa to Iceland, migrate thousands of kilometres to the Sargasso Sea in the Bermuda Triangle. This was discovered through attaching satellite tags to eels to track them and map their travel routes.

从北非到冰岛，所有欧洲的鳗鱼集体迁徙数千千米，到百慕大三角的马尾藻海。通过给鳗鱼装上卫星标签，人们才发现并定位了它们的游行轨迹。

Aristotle wrote the earliest known inquiry into the life cycle of eels. He speculated that they were born of "earth worms", which he believed were formed from mud, growing from the "guts of wet soil".

在人们已知的材料中，亚里士多德最早地记录了鳗鱼的生命周期。他推测鳗鱼是由"蚯蚓"演变的，他还认为蚯蚓是由泥巴形成的，并且在潮湿的泥土里长大。

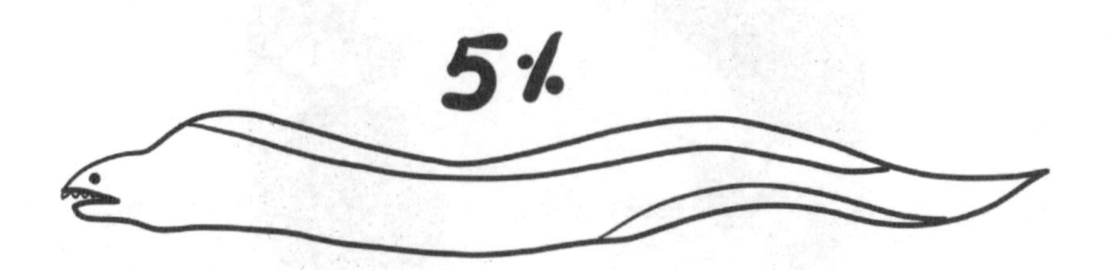

Illegal trade in eels is the largest wildlife crime in Europe. It is the European equivalent of the illegal ivory trade. Eels once represented 50% of fish biomass in European fresh water. The eel population has dropped by 95%.

非法买卖鳗鱼是欧洲历史上最大规模的野生动物犯罪，相当于欧洲的象牙非法交易规模。鳗鱼曾经占有欧洲淡水鱼类生物量的 50%，但是现在鳗鱼的总数已经下降了 95%。

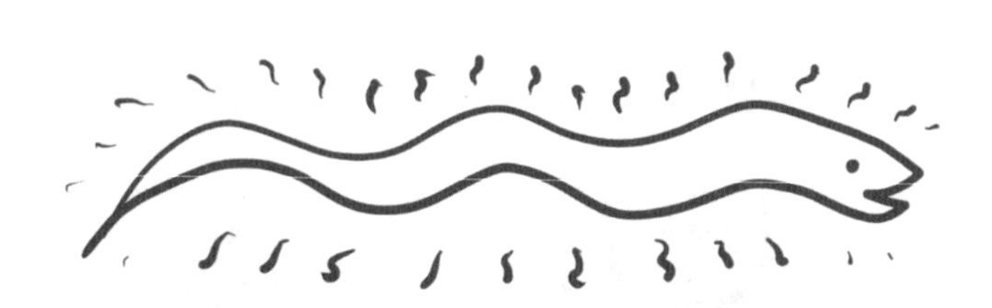

In July of every year, some mature individuals migrate back towards the sea, and travel up to 6,000 km to breeding grounds. As they enter the ocean, their gut dissolves, making feeding impossible, so that they have to rely solely on stored energy.

每年七月，一些成年鳗鱼会回到海洋中，经过长达 6 000千米的路程到达繁殖地。当它们进入海洋时，它们的肠道会溶解，以至于无法进食，于是它们不得不仅仅依赖自身储存的能量。

Before migrating, an eel's eyes enlarge for optimal vision in the dim blue ocean light. Its body turns a silvery shade to create a countershading pattern, which makes it difficult for predators in the open ocean to see it.

在迁徙之前，鳗鱼的眼睛会扩大以便在昏暗的海洋蓝光中获得最佳的视觉效果。它的身体会变得如同银色的阴影，来制造出一种伪装色，这让捕猎者很难在开阔的海洋里看到它。

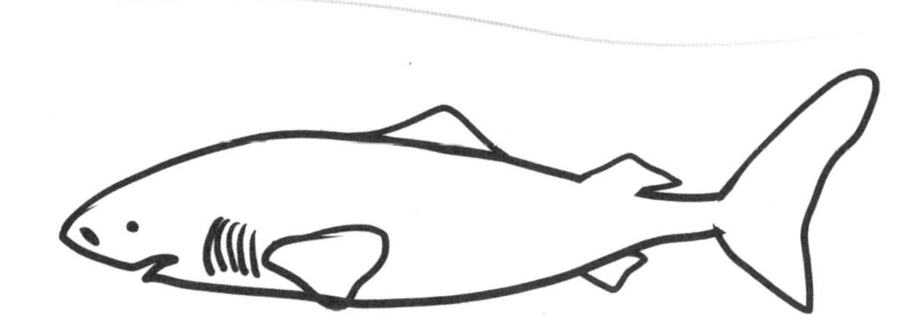

Greenland sharks, that can reach ages of 400 to 500 years, are the longest living vertebrates. To conserve energy in extreme cold, they swim very slowly and ambush their prey, the sea lion, while the sea lions are asleep. These sharks are born alive and do not hatch from eggs.

格陵兰睡鲨可以活 400 到 500 年，是最长寿的脊椎动物。为了在极寒天气里保存体力，它们会游得很慢，并且伏击正在沉睡的海狮作为猎物。这种鲨鱼是直接出生的，不用从卵中孵化。

Do you know who you are, and do you know what you want?

你知道自己是谁吗？你知道自己想要什么吗？

What do you think of a childhood that lasts a hundred years?

你觉得长达百年的童年怎么样？

Could you learn to swim like an eel?

你可以学习鳗鱼的游泳姿势吗？

Would you mind if you were called by the wrong name?

你会介意别人把你的名字叫错吗？

We need to find out why the eels are dying out. We can imagine several possible reasons for this: could it be overfishing, or too much pollution? Is there perhaps some disease present, or a specific parasite that has affected the eel population? Ask some people what they consider the cause of the decline, and then compare their opinions with the facts that appear in scientific literature. Once you have compared these facts with their opinions, communicate your conclusions to friends and family members.

我们需要找出鳗鱼濒临灭绝的原因。我们可以想象出几个可能的因素。可能是过度捕捞，或者是污染太严重吗？可能是新型的疾病，或者某种特别的寄生虫影响了鳗鱼数量吗？你可以询问一下人们，他们觉得造成鳗鱼总数下滑的罪魁祸首是什么，然后将人们所想出的原因与科学文献中的事实进行对比。对比过后，请将你的总结分享给你的朋友和家人。

学科知识

Academic Knowledge

生物学	格陵兰睡鲨是一种顶级掠食者；鳗鱼受线形寄生虫的影响；淡水鳗鱼有十分繁密的血管网，可以直接从空气或水中吸收氧气进入血液；鳗鱼在很深的地方产卵，产卵后它们就会死去；绿鳗鱼的黏液中含有海藻；水獭、麻鳽、苍鹭很喜爱鳗鱼，因为鳗鱼的脂肪含量很高。
化　学	电子产品中的电池可以通过内部氧化还原反应产生电压，这种电池没有生物相容性，某些生物体内有能够分离正负离子的生物电源；鳗鱼的放电细胞含有能够泵出钠离子的离子通道，在细胞两端产生相对电位差；电鳗串联起细胞，能产生600伏的电压。
物　理	鲨鱼用含氮废物来提升浮力，同时减小渗透压；产生推力和阻力的力；无肢运动有更高的能效；雷诺数是惯性力和黏性力的比值。
工程学	波动运动，也叫波浪式运动；表面的黏液使鳗鱼容易地从淡水转移到盐水；雷诺数适用于管内液体流动，飞机机翼上的空气通道与之类似；在鳗鱼身上装卫星标签来追踪它们的迁徙路径。
经济学	由于鳗鱼从濒危状态转为极度濒危状态，鳗鱼的单价上升；发展鱼菜共生系统的动力；利用激素人工孕育鳗鱼将成为一项跨国产业。
伦理学	人怎么能捕捞鳗鱼，却一点都不了解它们的生殖过程，或者设计鳗鱼人工繁殖，却没有完全掌握这种特殊鱼类的生长过程；在欧洲，鳗鱼非法走私规模相当于非法交易象牙的规模；为了增加鳟鱼的捕捞量，新西兰的殖民者曾试图消灭鳗鱼。
历　史	南太平洋岛屿的椰子树传说将鳗鱼和第一颗椰子联系起来；夏威夷的长鳗传说；鳗鱼曾是18世纪英格兰穷人的主食。
地　理	北大西洋的马尾藻海以4股水流为界，形成了一个海洋环流；日本幼鳗在马里亚纳海沟西侧孕育，那里邻近关岛；鳗鱼在海洋中孕育（有较多掠食者），在淡水中生长（较少的食物），这点与三文鱼相反。
数　学	雷诺数被用于计算层流在什么情况下转变为湍流。
生活方式	河鳗富含维生素和矿物质，所以为了均衡营养，可将它当作补充性食物；忙碌的生活节奏让人们没有时间观察周边事物，也无法了解新奇事物。
社会学	“滑得像鳗鱼”形容一个人狡猾或难以捉摸；某些泳姿的竞技规则要求选手只能在最初的和每个转身后的15米内，在水下使用海豚式打腿。
心理学	鳗鱼象征着活力、诱惑，以及隐藏的欲望；解析梦境：鳗鱼象征着丢失财富，使其从手中溜走；放慢生活节奏，用心观察，这样可以减小压力和增强信心；永远保持童心，一直准备好去玩、去探索、去发现。
系统论	由于水产养殖不受控，线虫已经遍布日本和欧洲的鳗鱼；鳗鱼已经在污染水域生活了近几十年，并将毒素带入食物链；鳗鱼的障碍识别能力和三文鱼一样，就像安了可以帮助它们迁徙的梯子似的。

情感智慧

Emotional Intelligence

鲨　鱼

鲨鱼很好奇鳗鱼急急忙忙的原因，他很在意生活质量，同时分享了他的担心。他阐述了自己对鳗鱼的同情之心，并向鳗鱼提供了简单的建议。他将电鱼和鳗鱼弄混了，并好奇人们为何误取名字进而造成混淆。他开诚布公，很坦诚地告诉鳗鱼，她的样子很吓人。随着对话展开，他发现了新的事实。他有一颗好奇心，而且当他不明白某事时，他会刨根问底。他表达了对鳗鱼坚韧性格的赞美，这增强了鳗鱼的自信心。

鳗　鱼

鳗鱼解释了为什么她没有时间可浪费，因为鳗鱼在迁徙时不能进食，只能靠身体里储存的脂肪生存。她强烈要求将自己和电鱼区分开，并阐明她的族群的特殊性，以此避免混淆。她也希望澄清关于自己呼吸方式的事实。她希望人们了解她的族群。她展示了自己的性格，及其族群的决心——坚持到达目的地，通过合作跨越一道道难关。她分享了她的智慧：知道自己是谁很重要，知道自己去哪也是。

艺术

The Arts

鳗鱼运用波形运动制造出数字8，这使鳗鱼轻松地游泳。数字8同时还是一种凯尔特符号，代表了团结、爱、永恒。凯尔特人曾经居住在淡水附近，那里有许多鳗鱼，同时，这也造就了他们的饮食习惯。为了在珠宝、腰带和领子上体现出这个数字，人们会把粗绳扭转弯曲出编织曲线，来表示无限。所以让我们一起编织粗绳，来学习如何呈现大自然和传统文化中的数字8。为了制作一个20厘米长的领子，你需要十倍长度的粗绳。

思维拓展

Systems: Making the Connections

人们已经食用鳗鱼上千年，但是鳗鱼仍然神秘。鳗鱼为了繁殖从河流游到海洋，甚至为了到达流入大海的河溪而穿越陆地。在一些欧洲淡水流域，鳗鱼占有了鱼类生物量的50%。为了到达大西洋或者太平洋中心的水流交汇处，鳗鱼会穿行几千千米。在人类看不见的最深处，每条雌性鳗鱼能够产出千万颗卵，然后死去。这些幼鱼会经过很长时间的发育，之后才会成熟。作为蛋白质的来源，人们对鳗鱼的需求不断增加。它们富含矿物质，而且被当作美味佳肴，这些都导致过度捕捞进而造成鳗鱼储量不足。过度捕捞带来污染，并破坏了鳗鱼的栖息地，使鳗鱼数量大大减少。但是，对鳗鱼的需求量依然很高。传统的鳗鱼养殖方式已无法满足需求。这推动科学家们探索鳗鱼繁殖的新领域。对鳗鱼的研究还仅仅是小规模的，但这些研究都具有开拓性。似乎鳗鱼将会带着它们的秘密进入坟墓，因为欧洲鳗鱼已经被列入国际自然保护联盟红色名录的极度濒危名单中。人们对于动物在全球的大迁徙很着迷，许多生物的迁徙过程都被完整地记录并研究，但是我们始终没有捕捉到鳗鱼迁徙的奥秘。几个世纪以来，它们的目的地一直没变过：海洋的某个地方，洋流在那里汇合，而且没有陆地，也没有任何国界线。虽然鳗鱼并不是最招人喜爱的动物，但是作为珍贵的蛋白质来源，它值得我们尊敬、赞美，以及保护。在这样的背景下，神秘的鳗鱼应该被研究、被理解，并且被分享。

动手能力

Capacity to Implement

鳗鱼很重要，因为它能平衡自然生态。让人担心的是，在过去的25年里，鳗鱼幼鱼数量已经下降了95%。所以让我们看看在阻止非法买卖象牙和鳗鱼的过程中，非政府组织和政府采取的行动有什么值得学习的地方？我们可不可以思考出平衡鳗鱼供需的方法？我们也需要进一步研究鳗鱼的生活习性和它们的繁殖方式，这是为了给这一物种创造一个安全的保护区。根据你的调研，你会采取哪些行动？

故事灵感来自

This Fable Is Inspired by

波琳·贾汉内特
Pauline Jéhannet

2010年，波琳·贾汉内特开始在家乡法国的昂热大学学习药理学。很快地，她将专业转为同一所大学的生态学、进化论和系统学，并且得到学士学位。她在瑞典的于默奥大学继续学习环境科学，钻研波罗的海的外来物种。她在荷兰莱顿大学得到了进化学、环保和生物多样性的硕士学位。之后她在荷兰吉滕贝格尔海洋研究项目、瓦格宁根大学战略与畜牧研究所、新西兰奥塔哥大学实践和学习。目前，她在瓦格宁根大学攻读博士学位。她的研究方向是通过研究鳗鱼排卵和早期幼鱼成长，以及它们最后成熟的一系列过程，更好地了解鳗鱼幼鱼的可控繁殖。

图书在版编目(CIP)数据

冈特生态童书.第七辑：全36册：汉英对照 /
(比)冈特·鲍利著;(哥伦)凯瑟琳娜·巴赫绘 ;
何家振等译. —上海：上海远东出版社，2020
ISBN 978-7-5476-1671-0

Ⅰ.①冈… Ⅱ.①冈… ②凯… ③何… Ⅲ.①生态
环境-环境保护-儿童读物—汉英 Ⅳ.①X171.1-49

中国版本图书馆CIP数据核字(2020)第236911号

策　　划　张　蓉
责任编辑　祁东城
封面设计　魏　来　李　廉

冈特生态童书
滑溜溜鳗鱼
[比]冈特·鲍利　著
[哥伦]凯瑟琳娜·巴赫　绘
靳维筠　译